Sea Creatures

Editorial Team Michelle O'Regan, Margaret Just, Alan Warburton

Illustrations Greg Grace

Design Anne Wilson

ISBN 978 1 9213070 7 2

Unit 2, 1 Union Street, Stepney, South Australia, 5069

Enquiries should be made to the publisher.

DATA FOR LIFE

www.axiompublishing.com.au

Printed in Malaysia, Reprinted 2009

Contents

What a wonderful and mysterious world exists below the surface of the sea!

Many creatures including mammals, crustaceans, molluscs and fish call the ocean, home.

Apart from the problem of pollution, the sea is a wonderfully balanced environment.

Predatory fish with their sleek body shapes swim effortlessly through the water, and the non-swimmers like crabs, lobsters and oysters rely on their tough shells for protection from predators as they travel the ocean floor.

Others rely on camouflage to blend in with their habitat, and some have venomous spikes to put off attackers. Coral, sponges, barnacles and tube-worms attach to the ocean floor or a rock and catch tasty morsels as they drift past.

How deep is the deepest sea? Do we really know what is down there?

The ocean is divided into 4 layers, depending on the amount of sunlight received.

1) Sunlit Zone surface–200 m (660 ft)

2) Twilight Zone 200–1,000 m (660–3,300 ft)

3) Midnight Zone 1,000–4,000 m (3,300–13,200 ft)

4) Abyssal Zone 4,000–6,000 m (13,200–19,800 ft) with depths far exceeding this in places.

A scientist who studies fishes is called an

Ichythyologist

Fish or fishes? A group of fish of the same species are called fish. However, two or more species of fish are called fishes. A fishy business!

For other words about sea creatures, go to the glossary.

As you would expect, most of the ocean life occurs in the sunlit zone. As you go deeper, the water is colder and there is more pressure on the animals' bodies. Some fish, sea jellies, octopuses and sperm whales will dive into the lower zones for food, but as no sunlight reaches the lower levels, no plants grow. In 2005 tiny organisms called ***foraminifera*** were discovered living in earth taken from the ocean floor at a depth of 11 km. At that depth the pressure is so great it's like having 50 jumbo jets piled on top of you!

There are many sea creatures yet to be discovered.

What are these "fishy" parts?

Gills

The fish uses gills to extract oxygen. They are adapted to take oxygen from water, as lungs do from air.

Gill cover (operculum)

This is the flap of bony plates and tissue covering the gill cavity.

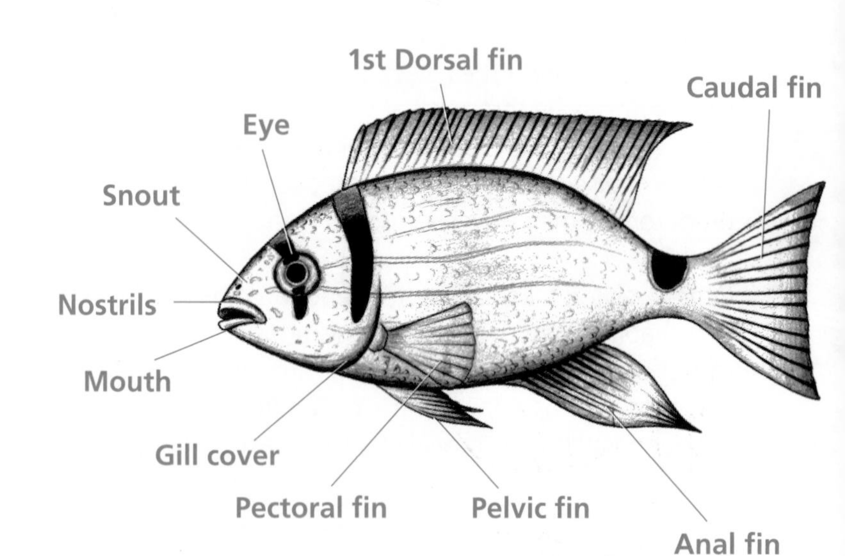

Nostril

These are used only for scent, not for breathing.

Lateral line

Pressure-sensitive receptors found in small canals along sides of fish, which help fish navigate in the dark.

Fins

Are the "wings and tail" mechanisms used by fish to swim and manoeuvre.

- ***Caudal*** (tail) fin is the main thruster in swimming.
- ***Pelvic*** (under the belly) fins help the fish to balance, turn and brake.
- ***Dorsal*** and anal fins (back and behind) are used to steer and stabilise and help the fish to remain upright when turning at high speeds.
- ***Pectoral*** fins (behind the eye) provide lift and balance.

Mouth shape

Just as birds' beaks are adapted to their diet, so are the shapes of fish's mouths adapted to their diet.

A) Large mouth for swallowing or tearing large prey.

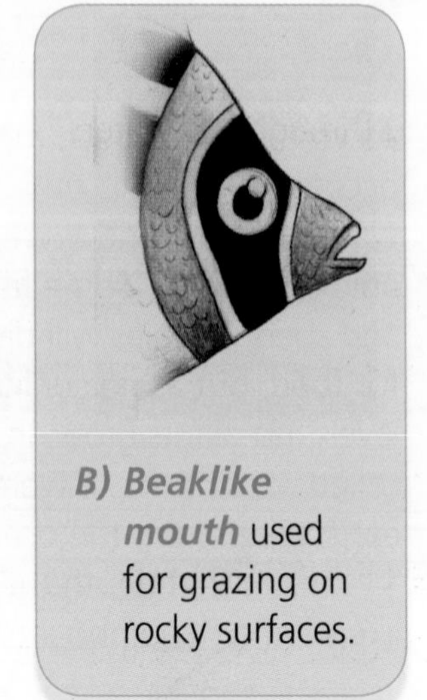

B) Beaklike mouth used for grazing on rocky surfaces.

C) Downward facing mouths are good for sucking up food from the bottom of the sea.

D) Long, skinny bills are great for poking into crevices.

Body shape

These relate directly to the fish's lifestyle:

Streamlined are the fast swimmers whose shape helps to lower water resistance.

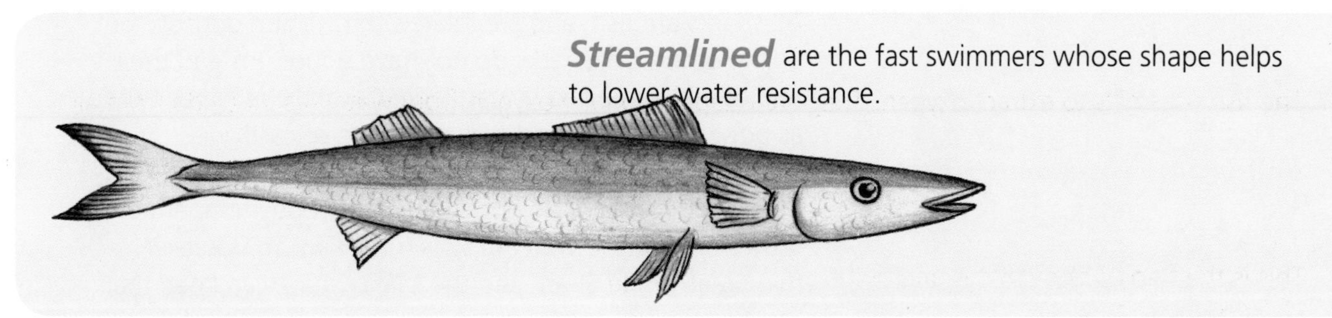

Elongated helps in manoeuvring into and through crevices. They often live in narrow spaces in coral reefs or in rocks.

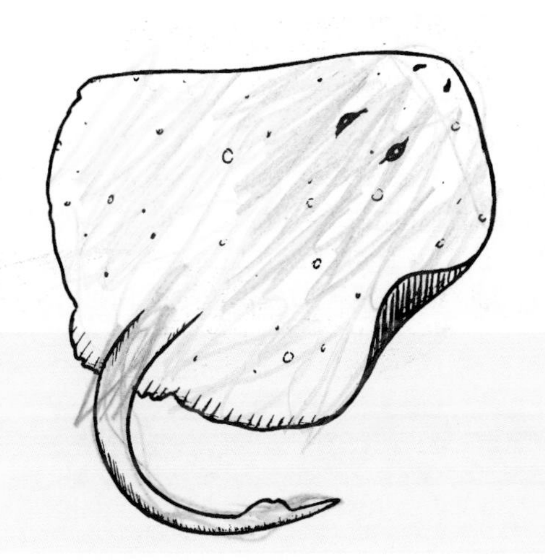

Broad and flat are the bottom feeders, adapted to living on or below the sand. Example – flounder.

Tall, thin shape is handy for entering vertical crevices and good for leisurely swimming but can accelerate with bursts of speed.

Underwater adaptations

Speed – Fish like Sharks and Mackerels use speed for catching their food and for escaping bigger, predators. These are some of the features helping them "fly" through the water:

- Smooth, cigar-shaped outlines, and pointed noses. These fish are streamlined!
- Strong muscles, powerful fins and tails for speeding through the water.
- Fins to give them good balance and fast turnarounds.

Camouflage and mimicry – Some fish don't use speed as a defensive or offensive strategy. Instead they have stealth and cunning. They melt into their surroundings and avoid being seen by the fish chasing them, or prey they are stalking.

- Colour and patterns, for example the Moray Eel.
- Texture, for example the Stonefish looks surprisingly like a stone.
- Very slow movement, for example the Leafy Sea Dragon, looks like a bulky bundle of seaweed.

Are sharks fish?

All sharks are fish, and are related to rays and dogfish, but unlike other fish, sharks do not have bones. Instead their bodies are supported by cartilage. Cartilage is not as brittle as bone. Our noses and ears are made of cartilage.

Most sharks are cold-blooded and their body temperature is the same as the water in which they swim. However the Great White Shark and the Mako Shark can keep their bodies at a warmer temperature. This is an advantage, because warmer fish swim faster.

Most ocean sharks must swim constantly to get enough oxygen, but a few species, such as the Tawny Nurse Shark, Whitetip Reef Shark, Lemon Shark and Nursehound Shark can all rest on the seabed and continue to breathe by pumping water over their gills with their fins.

	Shark Anatomy	Bony Fish Anatomy
Comparisons		
	Cartilage only	Bones and Cartilage
Swimming	Can only swim forward	Can swim forwards and backwards
Buoyancy	Large oily liver	Gas-filled swim bladder
Gills	Gill slits but no cover	Covered gill slits
Reproduction	Eggs fertilised in female body	Eggs usually fertilised in the water
Skin	Rough like sandpaper	Slippery, overlapping scales

Sharks have a clear **pecking order,** the bigger the shark, the more important it is.

Sharks use **body language** to communicate. Many species show behavioural patterns to intimidate … they arch their backs, point their pectoral fins down and swim stiffly. If this is not understood then a bite to the side of the head makes the point clearer!

Sharks like to have their own **personal territory.** The Blacktip Reef Sharks let others know they are trespassing by moving their jaw or opening their mouth.

Not all sharks are the same

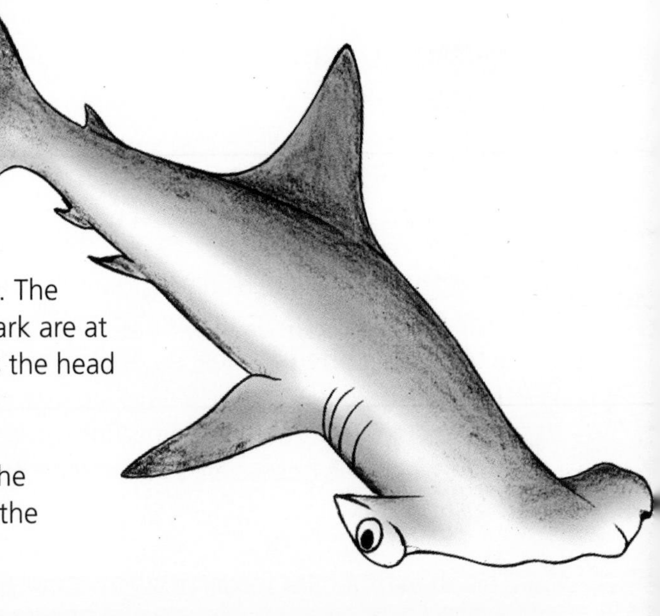

Hammerhead Sharks, like many other shark species, will drown if they stop swimming. Hammerheads swim in shoals. The larger sharks force their way into the centre, which happens to be the best position for finding a mate. At night they break off to search for food, or pair off for mating. Females often carry the scars where the male held her with his teeth. The shoal reforms at daybreak. The eyes of the hammerhead shark are at each end of its very wide head. As it swims along, it sweeps the head from side to side to gain the best view of its surroundings.

The hammerheads have their own butlers. They swim into the shallows where their friends, little Angel Fish, peck away at the parasites living on the shark's body.

Great White Sharks go through thousands of teeth during their lifetime. These are triangular, serrated, and razor-sharp. Their jaws can bite an elephant seal in half, bones and all.

The Great White can grow to over 6 m (20 ft) and are the largest hunting fish in the ocean. The Great White shark has a ferocious reputation as a man-eater. The Mako shark is also a fierce predator and dangerous to man.

More Shark Stories

Sharks of some kind have existed for 400 million years. Palaeontologists tell us they were around before the Dinosaur Age.

A shark's brain is tiny for its size but well-developed. They have good eyesight and see in colour. They also have special nerves to detect tiny electrical impulses emitted by the muscles of other fish.

Some sharks lay eggs, others have live young. After they are born, the babies swim solo. They instinctively fend for themselves, as sharks are not caring parents.

The Greenland Shark lives in very deep water and is the only shark known to survive these cold conditions. Because it lives deep in such cold, gloomy conditions, not much is known about its habits. It's pitch-black in the depths of the ocean - so a luminous parasite attached to the cornea of each eye acts as a beacon attracting prey for the shark to suck in.

The huge Basking Sharks such as the Whale Shark are not meat-eaters like most other sharks. They feed on plankton and krill just as the larger whales do.

What is the largest shark and the biggest fish?

The **Whale Shark** is both the biggest fish and the largest shark, but don't be fooled, it's *NOT* a whale. Whales are mammals. They have blowholes and must come to the surface to breathe.

The Whale Shark grows to 18 m (60 ft) and weighs in excess of 15 tons. Its feeding grounds are tropical waters. Despite its awesome size, it's a solitary, harmless creature. Scuba divers often clamber all over its bulk without the fish reacting badly. Whale sharks are lazy customers, which gets them into trouble! Sailing ships can accidentally ram them when the huge creatures are swimming just beneath the surface.

This shark is a filter feeder. It swims slowly, mouth open, sucking in huge amounts of water filled with plankton and krill. The water is filtered through the spongy tissue between its 5 large gill openings. Once the mouth is closed, the catch is sifted by gill-rakers – thousands of bristles in the shark's mouth. The gill-rakers save the food, then the waste is expelled through the gill openings. The catch is mainly krill and plankton, with occasional special treats of sardines, anchovies and squid.

This fish can process more than 7,000 litres (1,500 gallons) of seawater every hour.

Unlike other sharks which move only their tails, the huge Whale Shark slowly pulsates its whole body from side to side, travelling at the equivalent of a slow pace, similar to a human walking pace. The females give birth to hundreds of "pups" all around 60 cm (2 ft) long. New-born pups must fend for themselves, right the start.

Other Record Breakers

- **Longest known migration: Blue Shark**
  Up to 3,100 km (1,930 mi) across the North Atlantic Ocean.
- **Biggest meat eating shark: Great White Shark**
  It's reported that the biggest was caught off the coast of Cuba in 1945 and was 6.4 m (21 ft) long.
- **Fastest swimming sharks: Mako Shark**
  Up to 96 kph (60 mph).
- **Widest tailed shark: Thresher Shark**
  The width of its forked tail is almost as long as its body.

Which fish "fly" through the water?

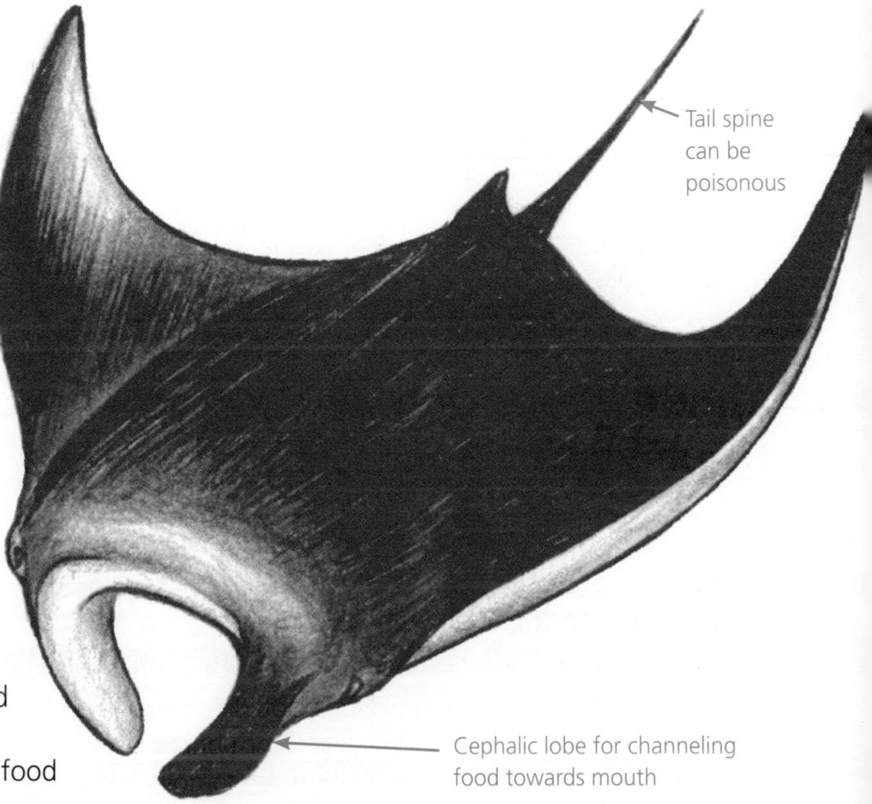

Manta Rays are the graceful giants of the ocean. Their large pectoral fins flap as they "fly" elegantly through the water. Some rays have sharp tail spines, and deadly poisonous dagger-like stings, but most are harmless. They have been known to leap out of the water like whales breaching. Maybe it is to follow their food source or perhaps it's a mating display. Scientists are not certain.

Tail spine can be poisonous

Awestruck divers who have seen huge squadrons of slow fluttering, silent rays, have said there is little in the ocean more wondrous.

Manta Rays have no teeth, so they sieve their diet of small fish, tiny crustaceans and microscopic plankton. Some use the large cephalic lobes above their eyes to channel food towards their mouth.

Cephalic lobe for channeling food towards mouth

These creatures grow to 9 m (30 ft) across and may weigh up to 1,350 kg (3,000 lb). They are generally dark brown on top with a white underside.

Did you know?

Sailors once called Manta Rays "devil fish" because of their "horns" (cephalic lobes) - these cunningly contrived plankton collectors funnel tiny sea dwellers towards their hungry mouths.

What is the fastest fish family?

The ***tuna family is the fastest fish family*** in the open ocean. They are in almost continuous motion. Their appetites are boundless and their prey of herrings, flying fish and squid are snapped and gulped down, as they speed through the water – doing what they do best!

Sailfish

No other bony fish range more widely across the open ocean than the tunas. Being heavier than water they must swim non-stop, or they would sink. They need to swim so fast to supply enough oxygen to their gills. The bone-hard sickle-shaped tails drive them forward with strong side-to-side strokes. Their massive swimming muscles contract and relax about 3 times more rapidly than those of any other fish.

The Sailfish can reach speeds of up to 110 kph (68 mph). Faster than the Mako Shark! That's almost above the speed limit!

What is an octopus?

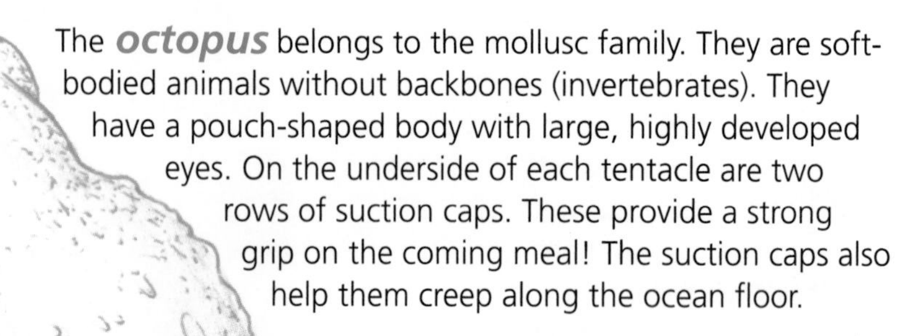

The ***octopus*** belongs to the mollusc family. They are soft-bodied animals without backbones (invertebrates). They have a pouch-shaped body with large, highly developed eyes. On the underside of each tentacle are two rows of suction caps. These provide a strong grip on the coming meal! The suction caps also help them creep along the ocean floor.

Should an octopus lose one of its tentacles it re-grows! Octopuses are solitary creatures and usually live alone. They move by jet propulsion, sucking in water then expelling from the back of their body. They confuse their predators by changing colour or shape. They can also squirt black ink into the water either to make their escape or to bamboozle their prey.

The word "Octopus" comes from the Greek, meaning eight feet or arms.

The Squeezy Octopus

- The body of an octopus has **no bones** and it's reported that an octopus can squeeze through an opening no bigger than its eye – a nasty surprise for a victim thinking it was safe!
- They **change colours** when they are angry or happy.
- They **keep their home** tidy by squirting all the shells and debris out after a meal.
- An octopus **can drill through a shell** with the tiny "teeth" on its tongue.
- The octopus has **3 hearts.** They are intelligent and can learn from each other.
- **Octopuses live** in all the world's oceans, but not in fresh water. The octopus has many nerves in the arms and suckers. It can actually taste with its suckers.

A mother octopus does not eat for 1–2 months while caring for her eggs. Her clutch size can reach 100,000 eggs! She cleans and aerates them until they hatch.

She also protects her eggs from fish and other predators. Without food for that long, poor mum octopus weakens and dies soon after the eggs hatch.

No Grandmother Octopuses!

Which sea creature has eyes as large as basketballs?

Amazing facts about the squid

- Squids have 2 hearts and 2 large eyes. The anatomy of squid eyes is similar to that of human eyes.
- The squid moves by a method of "jet propulsion", forcing water through flaps along its body.
- They are able to change the colour of their skin to blend in with their environment when hiding from predators.
- On Nov 2, 1878 the largest reported specimen of giant squid ran aground in Thimble Tickle Bay, Newfoundland. It weighed 2.2 tons, its body was 6 m (18 ft) long, and one of its tentacles was 10.6 m (35 ft) long.

There is much more to be learned about these enormous creatures, and the biggest is yet to be seen, no doubt!

There are no deep sea creatures more spectacular, than the

Colossal Squid and the Giant Squid.

The giant squid, as it's name implies, is a very large invertebrate, although it is thought the Colossal Squid, because of its broader body is the bigger of the two. Other deep-sea fishes are usually small, but squid and other invertebrates grow to huge proportions, lurking in the depths, below 180 m (600 ft).

Their eight muscular arms, as thick as a man's thigh, are studded with hundreds of suckers. In addition the Colossal Squid has hooks used for gripping its prey. Specimens with body lengths of 4 m (13 ft) have been discovered. In addition to the eight arms, are twin tentacles stretching to 15 m (50 ft). The flattened ends of these tentacles carry 100 or more suckers – used to bring food to its beak-like mouth, tearing the toughest flesh to shreds.

Little is known about these mammoths of the ocean, except they have fierce battles with Sperm Whales. Scars of squid suckers the size of dinner-plates have been found on whale bodies, and huge squid beaks have been found in whale stomachs.

The squid's arm can be as thick as a man's thigh

2 tentacles

What are whales?

Whales are easily the largest sea mammals. They have blowholes on top, through which they breathe when they surface. They have long, rounded bodies and almost hairless smooth skin. Whales are considered to be highly intelligent. Some migrate and others are territorial.

There are 2 types of whale:

1) **Toothed** whales such as orcas, sperm whales and belugas. Toothed whales are inclined to be quite social and often hunt in groups. Dolphins and porpoise are members of this family.

2) **Baleen** whales eat huge amounts of krill, plankton and other tiny creatures. They filter them from gulping seawater, through plates of baleen in their cavernous mouths. Baleen whales swim alone or in small groups.

The biggest and loudest of them all

The Blue Whale answers the question, "Which is the greatest?" On land or sea, the Blue Whale out-weighs the largest of the dinosaurs. It weighs as much as 30 elephants and can grow to 33 m (110 ft), which is the largest ever recorded, so far.

Producing 188 decibels of sound it is also the loudest. It is believed Blue Whales can hear each other underwater from 160 km (100 mi) away. Being a migratory animal Blue Whales don't bump into each other by accident, so their powerful sound is like an underwater mobile telephone.

Despite their huge size, their staple food is only tiny – plankton and krill, which they sieve through their baleen filters, just like the Baleen Whale. It has been estimated the Blue Whale eats up to 4 tons of food a day!

Did you know?

- **Females are larger than the males.** This applies to all members of the baleen family.
- The mother Blue Whale is **pregnant for 22 months,** longer than any other animal. Then she gives birth to a 250 kg (550 lbs) calf. What a whopper!
- After the birth, in the **7 months she nurses** her calf, the female Blue Whale loses up to 25% of her body weight!
- They have **two blowholes** for breathing.

Wondrous whale facts

Sperm Whales in Marguerite formation

The *Sperm Whale's head* is the biggest of any animal and accounts for a third of its length. Inside the head are 4 tons of an oil called spermaceti which makes the whale buoyant in the water. In the last century the Sperm Whale was hunted for its oil which was used for lighting in the home.

It is thought Sperm Whales rely on **echolocation** – a similar system to bats – to find food. Another theory is they use loud sounds to stun their prey.

If a whale is injured, the other members of the pod gather around in a **protective circle,** which is called the "Marguerite" formation.

The Sperm Whale's main source of food is the deep sea squid. It's estimated a Sperm Whale can eat up to a ton of squid a day.

Though it is a member of the white whale family, the Narwhal is coloured dappled green, black, grey and cream. The older the animal gets, the lighter in colour it becomes.

It is a noisy customer making loud clicks, squeaks, and whistles for both communication and navigation.

The male Narwhal has an extraordinary *spiralling "tusk "* or tooth, projecting 2–3 m from its upper jaw. Is it a jousting weapon, or do they use it as an ornament to attract females? Nobody really knows, but it definitely looks dangerous!

In the past, Narwhals were killed for the ivory in these tusks, which were sometimes passed off as the horn of the mythical unicorn.

Belugas – pure white whales – are nicknamed ***"sea canaries",*** because they make such a variety of sounds. Apart from chirping, whistling and clicking they clap their jaws together to make yet another type of noise.

The newborn Beluga calf is brownish-red in colour. After a year it turns grey, not becoming pure white until it's about 5 years old.

Unlike other whales, the Beluga has a *flexible neck.* It is able to turn its neck almost to right angles, handy for keeping track of mobile prey, such as squid.

Are there venomous sea creatures?

There are **more species of venomous fish than venomous land vertebrates,** including snakes. Researchers maintain there are over 1,200 species of venomous fish in our oceans, lakes and waterways. The venomous species include stonefish, catfish, lionfish, scorpion fish, toadfish, stargazers, and half a dozen other families. Most use their venom as a defence against predators.

Stonefish

Stonefish

The Stonefish is the most venomous fish known. It looks like a stone lying on the ocean bed. It uses its venom for protection, and when in danger it erects deadly spines along its back and fins. Every year people are badly injured by standing on Stonefish.

The Puffer fish carries so many toxins in its body that just a few mouthfuls are poisonous to humans. There is no known antidote, but in Japan Puffer fish is a delicacy, so chefs are especially trained to prepare the fish making it safe to eat. Maybe you would like to try something else on the menu!

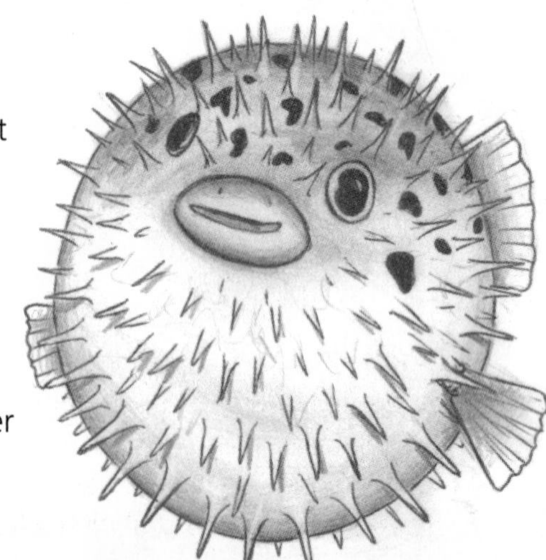

Puffer fish

Many ocean invertebrates contain venom – used either as a defence against attackers or for immobilising prey.

Blue-ringed octopus against golf-ball sized circle

The Blue-ringed Octopus is the size of a golf ball and found in shallow tropical water or in tide pools. It is browny-yellow with blue rings but when it's threatened, turns to yellow with bright blue rings. Even though the bite is painless, it injects a paralysing venom which is worse than any found on land. The Blue-ringed Octopus is very dangerous to humans, but most deadly for crabs, causing their body's systems to shut down.

The Box Jellyfish, also known as the Sea Wasp, are considered to be the most venomous of marine creatures. It is a transparent pale blue jellyfish with 4 distinct sides, with a single eye on each of the sides! Hanging from the "box" are tentacles up to 3 m long (10 ft) covered with 5,000 stinging cells which inject the deadly venom. The excruciating pain kills humans unless treated quickly.

Almost everything living in the sea is a predator of some kind, and part of "the food chain". In this way the ecosystem on both the land and under the ocean is finely balanced.

There are different ways to catch one's dinner:

- **Ambush predators.** The Flathead lies flat on the sandy bottom, looking like a brown stone. When a small fish passes within striking distance, the Flathead lunges at them, mouth wide-open, and that's that!
- **Lure predators** which actively dangle luminous bait to draw their curious prey to them, for example, the Anglerfish.
- **Swift predators** like the Mako shark which is the greyhound of speed amongst sharks. Barracudas are feared more than sharks in some parts of the world. They rip through a shoal of fish, slashing and tearing with their terrible teeth and later return to the carnage to feed.
- **Pack predators** such as Killer Whales and Dolphins. When they work as a team both these sea creatures become very efficient hunters, and woe betide their chosen victims.

Fierce predators

The **Great White Shark** reigns supreme amongst sea predators. It is the largest predatory fish in the ocean and is king of the food chain. Its favourite food is an unsuspecting seal or sea lion. The shark counts on the element of surprise, lurking beneath, then spearing upwards, clamping the seal, before breaching, and then splashing through the surface with its catch. The shark is not a chewer, but rips and tears. The sight of such an attack can be terrifying.

The Viperfish is one of the fiercest of the deep sea predators. It has an exceptionally large head and mouth with fang-like teeth. The fangs are so large they protrude outside the fish's mouth. It is thought the Viperfish swims at high speed towards its victim impaling it on its sharp teeth. Although only about 30 cm long (1 ft) it looks fearsome as its dorsal fins are tipped with light-producing organs called photophores.

Coral reef food chain

Sea life comes in all shapes, colours and sizes, and all of it is food. From the tiny plankton to the largest of whales, either living or dead, eventually it is eaten.

Animals must eat to survive, which often means eating a smaller animal. Plants and animals are connected to one another in intricate relationships called food chains or food webs.

A food chain links the predator and the prey. A food chain in the sea begins with plant life. The plants grow partly by converting energy from the sun, and also by using nutrients from the seabed.

Plants are the most abundant of all life forms and since they make their own food they are called ***"producers".*** The rest of the animals are ***"consumers"*** as they eat or consume plants or other animals.

How do reef fish outwit the enemy?

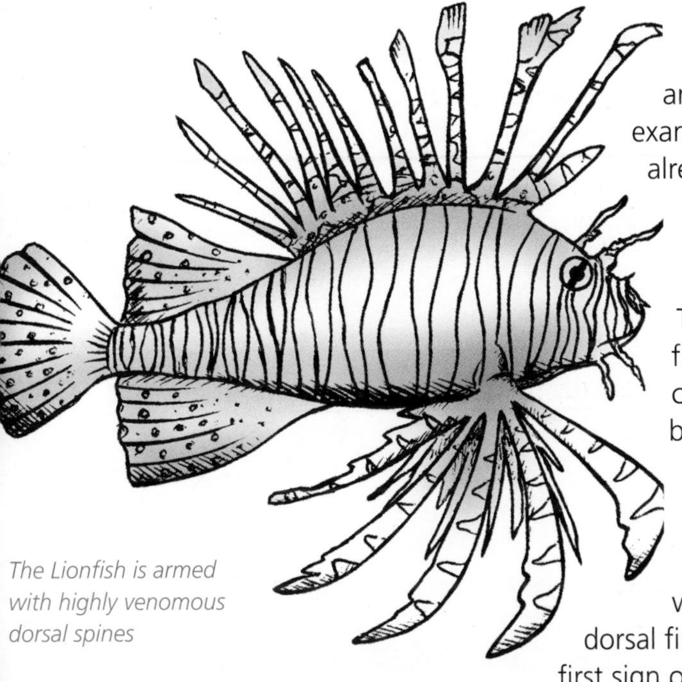

The Lionfish is armed with highly venomous dorsal spines

The superb hunters of the sea – Moray eels, sharks and barracudas – can still be outfoxed. Reef fish, for example, have evolved amazing forms of defence. We have already read about the poisonous fish.

Outer protection

To avoid the sharp, tearing teeth of predators, some fish have developed outer protection, often at the cost of mobility. Examples are the Trunkfish with bony armour, and Surgeonfish, named because it cuts aggressors with its razor-sharp spines.

The rough, leathery skin of the Triggerfish is a protection against attack, but it also comes armed with another defensive weapon. It has a 3-spined dorsal fin which can be locked into the upright position. At the first sign of danger, the Triggerfish races for a rocky crevice and wedges itself there with the help of its dorsal anchor.

Can you name 5 ways in which fish protect themselves from attack?

Un-tasty!

Some fish avoid death by being unpalatable. Species of Parrotfish secrete through their skin a loose-fitting, mucous-covered sleeping bag. It's definitely not too tasty, and is thought to hide the smell of the fish from predators. A number of reef fish are poisonous to man and likely poisonous to other fish. The Soapfish has a bitter soapy mucous on the skin which puts would-be attackers off their dinner.

Big spot on Butterfly fish used for disguise and deception

Camouflage

In the shallow, well-lit waters of the coral reef, some fish rely on their colour pattern to avoid attack. Camouflage means "colour that conceals". In the world under the sea, it is a strategy that works both in attack or defence.

Brightly coloured reef fish seem to be advertising their presence, but the colours act to break up the fish's outline against the multicoloured background of the reef itself.

Another type of camouflage is disruptive shading – contrasting bars or stripes and blotches that break up the fish outline.

Some fish are able to change their colouring. They have special "chromatophores" in their skin making this possible.

Trickery

Many reef fish rely on tricky behaviour to avoid being eaten. Brightly coloured Damselfish stay close to the tangles of coral branches and at the first sign of trouble, they seek shelter in the "forest".

What are dugongs and manatees?

These are sea mammals. They are big, peaceful creatures with plump, round bodies, two front flippers, and a large flat tail. The face is gentle looking with a wrinkled snout, whiskers and tiny eyes.

Dugongs and Manatees are the only sea mammals which are complete vegetarians. However because of their size they need to eat huge amounts to stay alive, about 32 kg (70 lb) of water plants every day!

Dugongs live along the coasts of New Guinea, Asia, Australia and East Africa. The 3 species of manatee are West Indian, West African and Amazonian. Unlike the Dugong which lives only in sea water, some Manatees are able to live in both sea water and freshwater. Freshwater Manatees sometimes use their flippers as legs when they are grazing in shallow water.

Fascinating facts

- **Dugongs have a nickname "sea pig"** because they like to root around for food on the seabed with their big bristly snouts.
- They can **hold their breath** underwater for about 20 minutes, and live for up to 60 years.
- Dugong **mothers have been sighted cradling their calves** with their flippers. This is because the mother's nipples are located under the flipper.
- Sailors once believed in **mermaids.** It's likely Manatees and Dugongs were imagined to be mermaids. No offence to the animals - but they were not pretty mermaids!

Can you work out which is the Dugong?

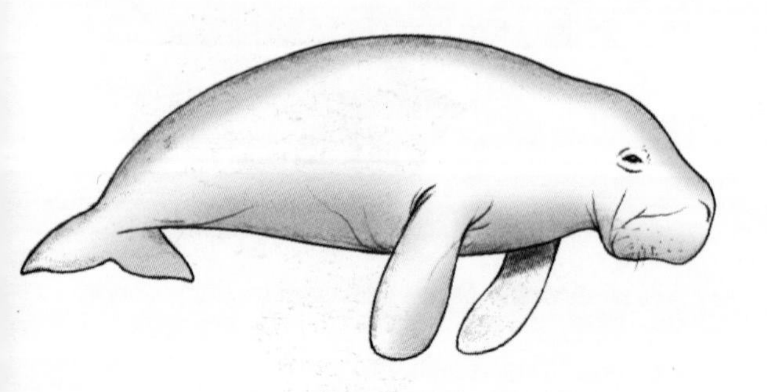

The main differences between Dugongs and Manatees:

- *Size:* A Manatee is slightly larger than a Dugong.
- *Tail:* The Manatee's tail is very rounded but the Dugong's is the shape of a whale's tail.
- *Tusks:* Mature male Dugongs have two teeth sticking out of their mouths like tusks, unlike Manatees which have none.
- *Fingernails:* Manatees have small "fingernails" on the end of their flippers which Dugongs don't have.

Which sea creature has ivory tusks and whiskers?

Walruses are known for their large size, ivory tusks and whiskers. Their cinnamon brown body has a rounded, all-in-one shape. The Atlantic Walrus is slightly smaller than the Pacific Walrus.

Their rough flippers are hairless. The pectoral flippers (ones at the front) have a similar bone structure to land mammals, even though they are shortened and modified. Each triangular shaped fore-flipper has five digits of equal length with a very small claw.

Although they have good hearing, they do not have visible ears, and their back flippers rotate under their bodies when they waddle around on land.

- **Both male and female walruses have tusks** used to help lift their body onto ice or rocky shores, as well as for fighting. Walruses are social animals, but the males and females form separate herds. When it comes to deciding who rules, the most aggressive animals with the longest tusks and largest body usually dominate.
- Walruses are **pinnipeds.** Pinna, meaning wing or fin and pedis, meaning foot. They eat molluscs (clams), worms, sea cucumbers, crustaceans and other soft-bodied animals.
- **An adult may eat between 3,000–6,000 clams at a single feeding.** They use their sensitive whiskers to detect food on the sea bed. They then crush the shells, suck out the body of the clam and swallow it whole. Delicious!

It's not all easy for the Walrus

Walruses have predators too. They are hunted by polar bears, killer whales and humans.

They have an air sac in their throat. It acts like a "floatie" so when it wishes, the walrus fills up the sac, to keep its head above water, and take a snooze.

Male walruses are **four times bigger** than the females.

Are polar bears good swimmers?

Polar Bears are powerful swimmers. They use their large front paws to propel themselves through the water and their back legs to steer. They dog-paddle with their head and much of their back above water. The layer of blubber under their skin helps them float, and they have been seen swimming 80 km (50 mi) away from any ice or land.

Polar Bears have rough pads on their feet which prevent them slipping and sliding on the ice. They live along shores and on sea ice in the bleak Arctic. When ice forms over the ocean, and the bears cannot fish, they head out onto the ice to hunt seals.

The bear waits silently by a seal's breathing hole for the seal to surface. As it breaks the water, the polar bear lunges and sinks its deadly teeth into the seal's head. In autumn, pregnant Polar Bears make their dens in snow banks. They stay put throughout the winter and give birth to 1–3 cubs. Did you know it's warmer inside the snow than outside? In springtime, the mother emerges with her brood. She nurses them for another two and a half years and in that time protects and teaches them how to hunt and fend for themselves.

Do you know why it's much colder outside a snow bank or igloo than inside?

Where do European and American eels go to breed?

No-one knows exactly where these freshwater eels spawn (lay their eggs), but the smallest American and European eel eggs have been found in saltwater near islands at the edge of the salty Sargasso Sea in the middle of the North Atlantic Ocean. After spawning, the adults die and the larvae drift across the oceans for thousands of miles, returning to the freshwater homes of their parents in Europe or America. There they grow and mature until years later they too return to the Sargasso Sea to spawn and die.

The colourful array and diversity of Eels

- The **Giant Moray eel** may grow to 3 m (10 ft) long. This is a family feared more than most eels, maybe because of their appearance. They have large mouths and fearsome teeth – so don't mess with a Moray! Fortunately they only attack humans when surprised, or threatened at close quarters.
- The **Zebra Moray eel** has distinctive black and white striping and powerful teeth which help it crack open crabs and other crustaceans. To back up its fearsome appearance, it's more aggressive than the larger cousin, the Moray.
- A fully grown 2.5 m (8 ft) long **Electric eel** can deliver enough of an electric shock to kill a man – up to 600 volts! Here's a surprise – in spite of its eel-like looks it's not really an eel, but a fish.
- The long snout of a **Snipe eel** looks like a bird's beak. Its large eyes help find prey in the gloomy depths.

The Zebra Moray Eel

How does a jellyfish stay afloat?

The Portuguese man-of-war has many venomous tentacles

The jellyfish's ***tissues are 95% water.*** Rhythmic contractions of the "bell" give the jellyfish enough of an upward push to keep it from sinking. It spends life drifting on the surface of the ocean.

The umbrella shaped bell is a thin, double-walled sac filled with a gelatinous substance. In the centre of the bell is a digestive cavity, and hanging from that is the creature's ***funnel-like mouth.***

Trailing beneath the bell are ***tentacles*** covered in nematocysts which shoot poisonous darts into unsuspecting small fish. Victims are hoisted into the jellyfish's mouth and digested.

No jellyfish is deadlier than the Sea Wasps of the tropical Pacific and Indian Oceans. Baby jellyfish start as buds growing out from an adult, then settle on the ocean floor growing into saucer-like larvae and gradually float to the surface.

The Portuguese Man-of-war has many venomous tentacles. The Portuguese Man-of-war is like a blue bubble, half the size of a football floating on the surface, with long stinging tentacles hanging beneath. The venom is 75% as poisonous as a cobra's, but the really nasty thing about this fellow – his tentacles can still sting long after he's dead.

What opens clams with its suckers?

The **Starfish** opens clams by fastening its suckers on both sides of a clam's shell and prying them apart with strong steady pressure. It is amazing to watch this performance, from a creature which looks barely alive!

Hundreds of tiny tube feet stick out from Starfish "rays". They are used for walking, gripping prey and probably for smell. Thanks to these feet, starfish have been measured speeding along at 30 cm (one foot) a minute!

There are about 1,800 different species of starfish and they are echinoderms – not fish at all!

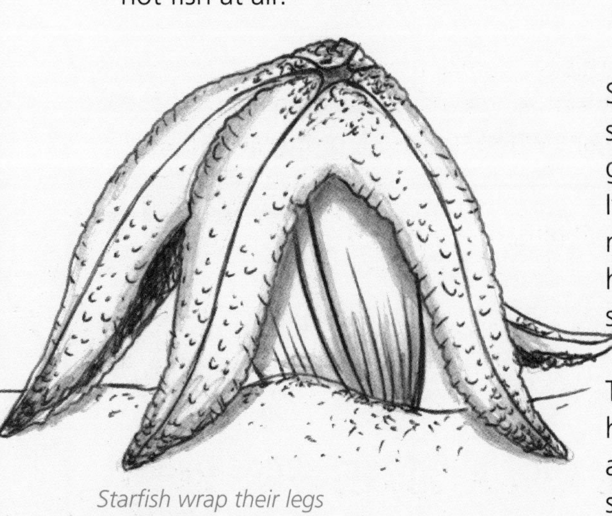

Starfish wrap their legs around clams and prise open the shell

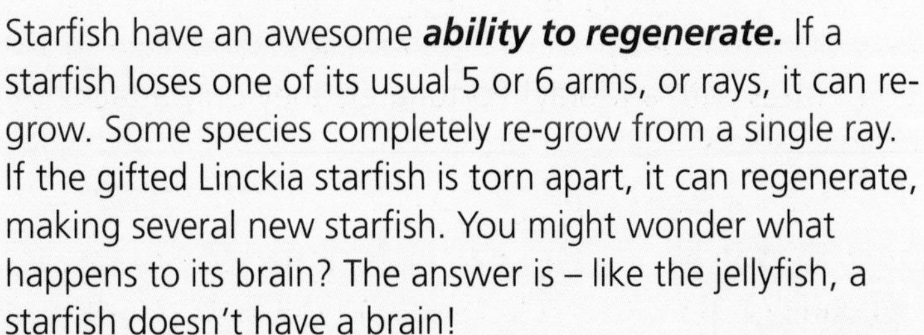

Starfish have an awesome ***ability to regenerate.*** If a starfish loses one of its usual 5 or 6 arms, or rays, it can re-grow. Some species completely re-grow from a single ray. If the gifted Linckia starfish is torn apart, it can regenerate, making several new starfish. You might wonder what happens to its brain? The answer is – like the jellyfish, a starfish doesn't have a brain!

The **Crown-of-thorns** starfish has 10–20 arms and a huge appetite for live hard coral. It is a coral reef killer and, as a result of this "bad behaviour", conservationists organise starfish hunts to remove them.

Sea turtles – reptiles that live in the sea

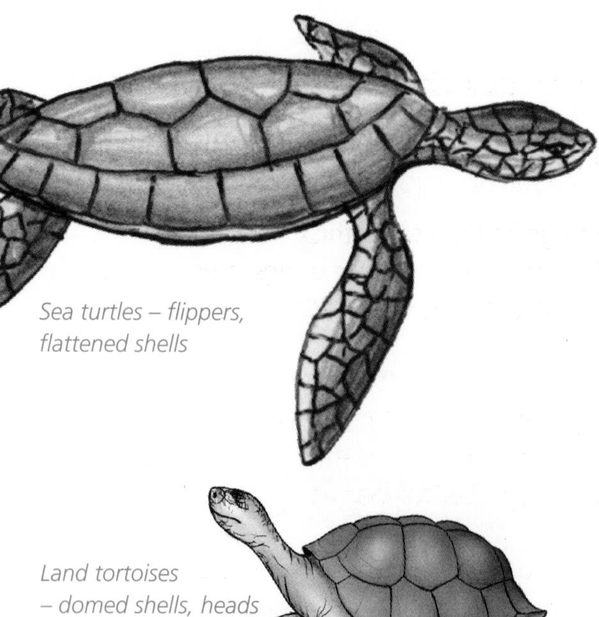

Sea turtles – flippers, flattened shells

Land tortoises – domed shells, heads that retract and clawed feet

Sea Turtles are reptiles which eat, sleep and live in the water, but return to shore to lay their eggs. Long ago, in the age of dinosaurs, turtles lived on land but, probably because it was safer, they adapted to a life in the ocean. There are several different types with interesting names such as Loggerhead, Hawksbill, Green, Black and Flatback.

Scientists believe Sea Turtles return to the same beach where their mothers, grandmothers and great-grandmothers laid their eggs. Before nesting time, sea turtles can be far away from their nesting beach, so they set off on a long journey. How do they find their way? Nobody knows, but some believe Sea Turtles are guided by the earth's magnetic field.

When it's *time to nest,* the female clambers onto the beach where the sand is dry. She starts digging her nest. By using her front flippers to clear away the loose sand on the surface, then her back flippers to dig a deep hole for the eggs.

She lays up to 100 eggs at a time, and often half a dozen times when nesting. Why so many? Well, some won't even hatch and the others will never reach adulthood because of the dangers to young turtles from predators. After laying them and burying her clutch, the mother returns to the sea. From that stage they are completely on their own!

Baby Sea Turtles hatch about 2 months after the eggs are laid. They are no bigger than a bar of soap. It takes about 3 days to dig their way out of the nest and head for sea. The hatchlings know instinctively to wait for the cover of night before making the distance.

If the newly hatched turtles avoid the dangers – hungry crabs and birds – they enter the sea and swim all night and into the next day, until they are many miles from shore. Here there are fewer enemies and they begin feeding on plankton, which floats close to the surface. Until their shell hardens they are vulnerable, so they hide in floating seaweed. Scientists say there is still much to learn about Sea Turtles.

Did you know?

- The Leatherback Turtle can grow as big as a bathtub and does not have a hard shell like other turtles but is covered by leathery skin.
- Unlike land turtles, Sea Turtles can't pull their heads back into their shells.

Introducing the bird swimmer

Penguins are flightless birds with black and white waterproof feathers.

Although they cannot fly, their streamlined body is beautifully designed for swimming. They spend most of their time in the water, chasing and catching fish.

All penguins have a large head, a short thick neck on an oval-shaped body with two flipper-like wings, two webbed feet and a stubby tail. Like all birds, they cannot breathe underwater.

Penguins have a layer of fat (blubber) under their skin, helping to keep them warm. In cold weather flocks of up to 5,000 penguins huddle together for warmth.

Penguins are viewed as food by Leopard Seals, Sea Lions and Killer Whales. The Australian Sea Eagle and Skua (a large sea bird) also include penguins, or their chicks, in their diet.

Penguins seem to enjoy water sports like diving and surfing. On the ice, they hop and waddle along, but are fond of **snow-belly tobogganing.** Using their short wings partly for balance and support, penguins can belly toboggan along frozen flat ground for a surprisingly long time.

Considering the penguin's short "legs", sliding on the belly sure beats walking!

Adelie Penguin diving from cliff face

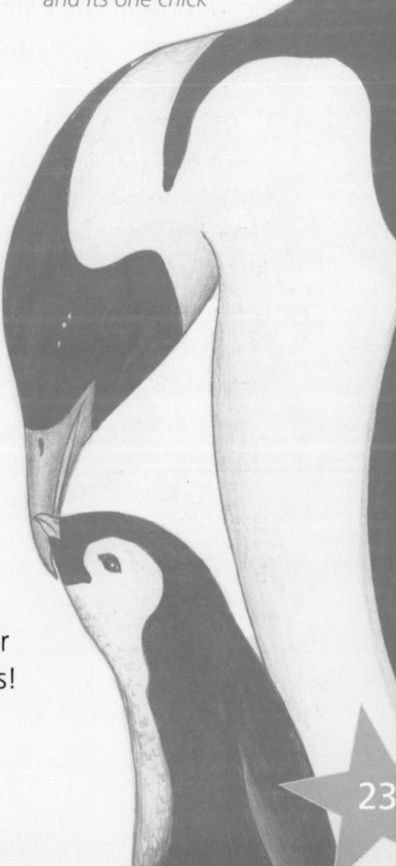

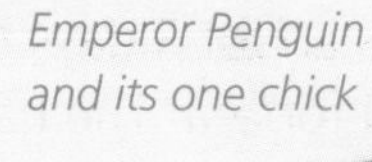

Emperor Penguin and its one chick

There are 17 different penguin species, and all live in the Southern hemisphere, from the South Pole to coasts of South America, Australia, Africa, New Zealand and the Galapagos Islands.

Adelie Penguins are amazing athletes and have adapted well to life on and around the pack ice. They are able to catapult themselves out from the water to several times their own height onto an iceberg or ice floe. They also dive spectacularly from great heights, deep into the icy waters. Because their bones are solid, unlike birds that fly, they can withstand diving to great depths, 18 m (60 ft) or more!

Endearing Penguins

- ***Penguins are able to drink sea-water*** thanks to a special gland, which filters the salt out of the water.
- The male incubates the egg on his feet tucked under a warm feathery belly for nine whole weeks and doesn't eat, drink or go to the toilet during this process!
- The ***African Penguins,*** earn their nickname "Jackass Penguin" by braying loudly when they want to make a point.

What is a crustacean?

A crustacean is a creature with an **exoskeleton** – a very tough, hard outer-shell made of a horn-like substance called chiton. Most, but not all, live in the sea. To allow for mobility and flexibility this shell grows in sections. Crustaceans are distant cousins of insects and spiders. Some, such as yabbies, are freshwater creatures.

As the animal grows, its ***shell must be removed and discarded.*** While the new shell is hardening, the crustacean is at its most vulnerable. But they do have a few weapons to fall back on. Many crustaceans have fearsome claws which can apply bone-cracking pressure. The ***Snapping Shrimp*** stuns its prey with a loud "explosion" made by snapping a large claw.

They have ***5 or more pairs of jointed legs***, a pair of antennae and often a pair of "nippers". The body has a head, a middle and a tail.

Most are nocturnal, hiding under sand or in rocks by day and coming out to feed at night. They feed on plant and animal matter. They are eaten by triggerfish, squid and many other fish which have jaws or beaks able to break through crustacean shells.

The Ghost Crab is a night scavenger found on warm sandy beaches. Its colouring matches the beach sand. It can run at 15 kph (10 mph) and has eyes used as periscopes when burying itself into the sand. All these features make this elusive customer tricky to find and catch. So it's called the Ghost Crab!

For protection, the **Hermit Crab** takes over the shell of a dead winkle or snail. As it grows, it finds a larger shell to call home. Hermit crabs are known to squabble over the shell. The loser then has to go house hunting!

Ghost crab

Lobster

A 9 lb female lobster may carry as many as 100,000 eggs! For 9 to 12 months she carries them inside her body, and then a further 9 to 12 months attached to the swimmerets (swimming legs) under her body. When they hatch, the larvae float on the surface for 4–6 weeks, then fall to the bottom to develop further. Of 50,000 larvae, possibly only 2 will reach maturity.

Spiny Lobsters, also known as Crayfish, migrate in a bizarre fashion. After forming a single file, they crawl along, each clutching onto the tail of the one in front of them.

Crayfish make a loud rasping noise to deter predators by rubbing their antenna against a smooth part of the shell.

Lobsters are able to discard a limb allowing them to escape greater injury. This is a life-saving strategy. They are also able to regenerate some body parts, such as claws, walking legs and antennae. They have a primitive nervous system, missing the part of the brain that perceives pain, the cerebral cortex.

Why do crabs travel sideways?

The crab's leg joints are similar to our knees, except they face sideways not forwards. So crabs must travel sideways. Which is an advantage for the crab as it can claw, wriggle and squeeze sideways into narrow crevices.

What is the biggest crustacean?

The Japanese Spider crab – can measure up to 36 cm (14 in) across the body and have a claw span of 2.7 m (9 ft). One of these crabs was reported as weighing 6.3 kg (14 lbs) with a claw span of 3.6 m (12 ft).

Their natural habitat is the bottom of the Pacific Ocean around Japan. They usually live at depths of 200–300 m but must migrate to shallower waters around 50 m (160 ft) to lay their eggs. These monster crabs are believed to live to an age of 100 years!

What is the heaviest crustacean?

The ***Atlantic Lobster*** – there are several records of individual lobsters weighing more than 9 kg (30 lbs). The all-time record goes to "Mike" who was caught in 1934 and weighed 19.2 kg (42.3 lbs)! The female lobster mates when she has just moulted, and her shell is soft. The two lobsters "dance" with their claws closed.

Stay-at-home crustaceans

Barnacles are crustaceans, often mistaken for molluscs because of their shells and because they remain stationary. They are often found attached to piers, boats or rocks, secreting a special cement to glue themselves into position.

As larvae, barnacles are mobile, but stay in place once they find a suitable niche. They then squeeze cement from their antennules to form a shell around themselves.

When the tide is out, the shell closes, so they won't dry out. Under water, the shell opens and the barnacles flap their little legs to attract plankton, which is their food.

The **Pea Crab** is less than a centimetre (half an inch) across its shell. It's also called the Oyster Crab because it often lives in the shell with an oyster, like a boarder. This system - where two animals live closely together - is called a symbiotic relationship. The Pea Crab doesn't seem to harm the oyster, but feeds on the same food.

What lurks in the depths of the ocean?

Perhaps the most curious animals on Earth are those that have evolved in the world of darkness and sparse resources within the oceans depths. Many of these fish living in the abyssal zone are luminescent. Here lurk the Gulpers, Swallowers, Anglers and Suckers. The deep-sea floor is populated with spiny creatures and sea cucumbers.

Deep sea fish generally have common features:

- Large mouths and teeth
- Large eyes
- Large stomachs in proportion to their size
- Dark colouration
- Some form of lighting or light organs

Weird Deep Sea Wonders

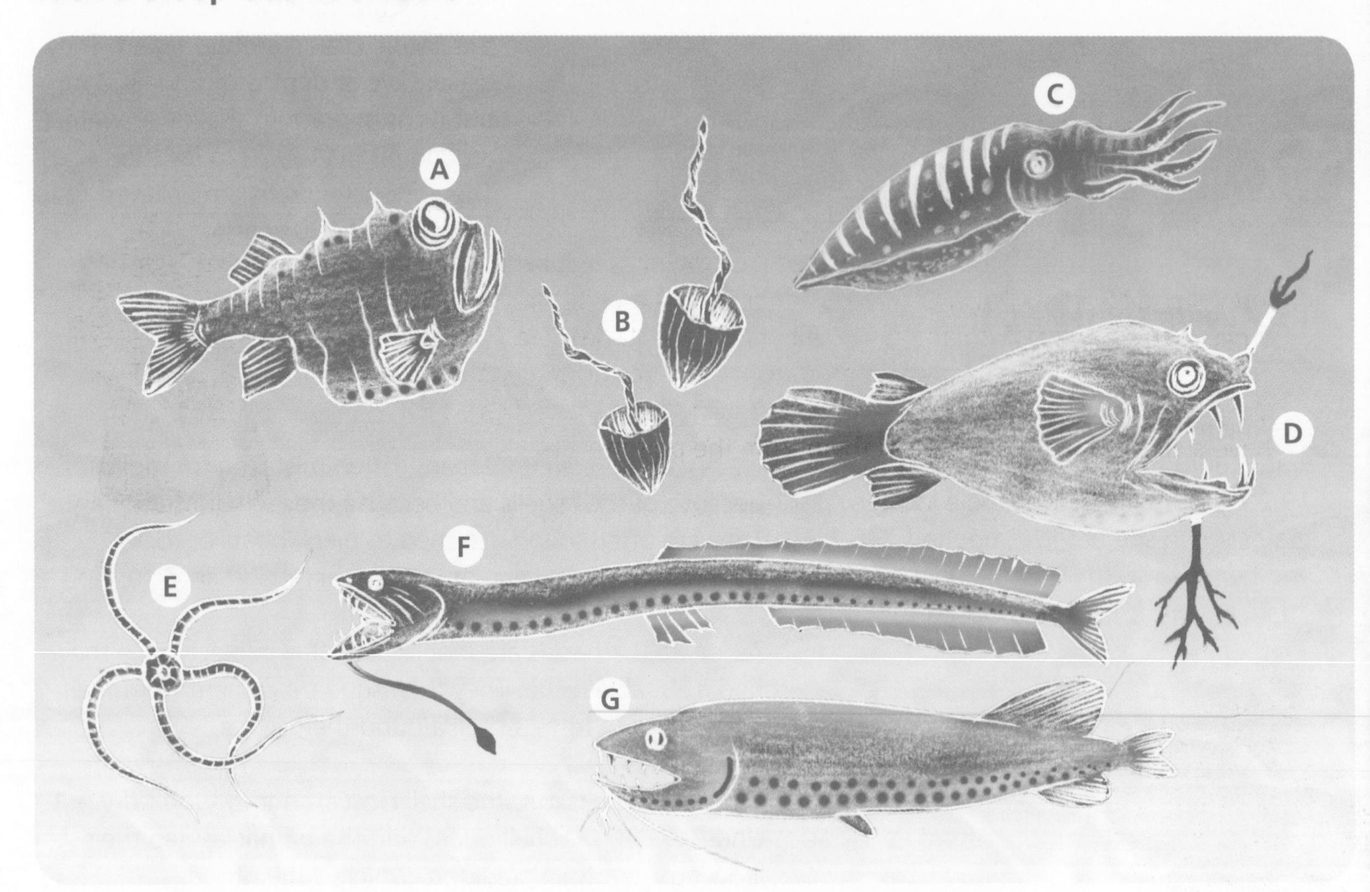

A *The Fangtooth fish or Ogrefish opens its huge mouth and just sucks another fish inside.*

B *Deep sea jelly fish have glowing tentacles to attract unwitting fish.*

C *The Cuttlefish has green blood, three hearts and is able to change colour in a flash. It doesn't only change to camouflage itself but lights up like a neon sign in red-orange and blue-green pulsing light.*

D *This Angler fish uses both the bulb that glows above its head and the hanging garden of glowing filaments below its jaw to lure prey into its ferocious jaws.*

E *The Brittle Star has five glowing arms that radiate from its body. It can live at great depths on the sea bottom.*

F *The Viper fish is a fierce hunter. It uses the barbel which hangs from its jaw to lure fish and its flanks are lit up as well. The mouth is full of very sharp teeth.*

G *The Stomiatoid is lit like an aeroplane, all along its flanks. It is another fierce predator of the murky depths.*

Molluscs have soft bodies, either inside or outside, a shell. They have no close relatives in the animal kingdom. The most complex molluscs are octopuses and squid where the "shell" is inside the body, forming a primitive skeleton. Molluscs which can't swim, walk, creep, shuffle or else stay in fixed position. These are Clams, Mussels, Oysters, Scallops and Razor Shells. The most remarkable mollusc is the Giant Clam, found in the Indian and South Pacific Oceans, growing to more than 1 m (3 ft) across and weighing more than 200 kg (440 lb). Should a diver get a foot stuck in a giant clam he is in serious trouble!

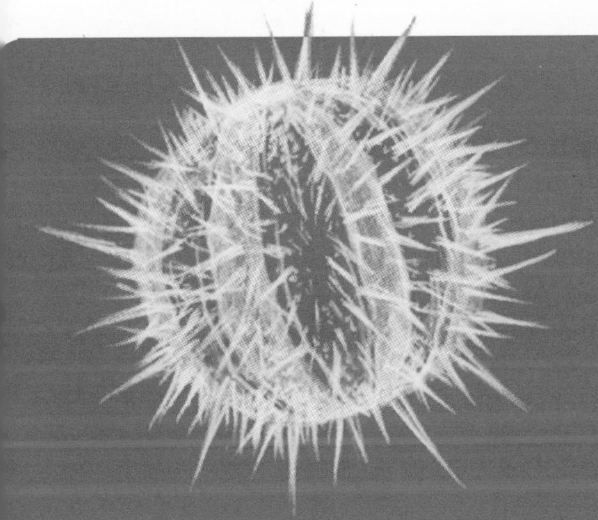

Sea Urchins look just like spiky balls, but among these bendy spikes are hidden tubed feet used for both travelling and collecting food. Underneath, Sea Urchins have mouths like claws. They creep along the sea bed contentedly munching away at tiny plants and animals. Like the Starfish and Sea Cucumber, the Sea Urchin doesn't have a brain.

Sea Anemones are flesh-eating sea creatures which look like flowers. They are related to jellyfish but instead of floating around, they stay put like a plant, and trap small creatures in their stinging tentacles. Sea Anemones secrete a poison in their fronds, protecting themselves from predatory fish. Despite this they are colonised by certain small fish like clown fish which have a special coating to protect them from the poison.

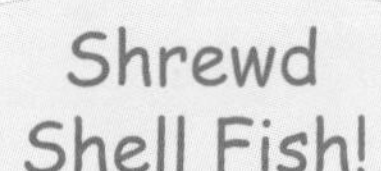

- **Mussels** have the ability to spin a fine strong thread called a byssus thread which "glues" the mussel in place. They use this thread to cling on to rocks, piers and other places.
- The **Scallop** uses its hinged shell to swim away from a predatory starfish. It claps the two halves of its shell together to make the water squirt out in a jet, leaving the slowcoach starfish behind.
- The **Dog Whelk** is a flesh-eating sea snail. It eats mussels and limpets. It bores a hole in the victim's shell with a drill-like tooth, then poisons the mussel through the hole and sucks out the flesh. Finally it gets inside and scrapes out the soft insides. All in a day's work for the Dog Whelk!
- **Limpets** are incredible creatures! They "graze" over rocks during the day, feeding on tiny bits of food, then in the evening they always return to the same place to sleep. No one has worked out how they manage to find their way back to home base!

More fascinating facts about sea creatures

- The **Ocean Sunfish** also known as the Mola Mola, is the most massive bony fish in the world. It is a huge fish sometimes weighing more than two tons and appears to be nearly all head as its body is very broad and its fins are set well back. They are not related to sharks, but they are often mistaken for one as they swim close to the surface displaying their dorsal fin. The leathery skin of the Mola Mola is host to more parasites than any other sea creature.

- **Flatfish have both eyes on one side** of their head. Young flatfish, such as Plaice, have eyes on both sides of their head like most fish. When a few weeks old, one eye starts to shift across the top of the head. As this happens the fish descends to live on the floor of the sea. The adults lie flat on the bottom with both eyes facing upwards. They can rapidly hide themselves under the sand, with only their eyes uncovered.

- The **Coelocanth** (pronounced see-la-kanth) was thought by scientists to be extinct. No bones or fossils younger than 60 million years had been found. In 1938, Marjorie Latimer was in charge of a tiny museum in South Africa. She had befriended a sea captain who would often invite her to come and see his latest catch – just in case there was something of interest to her. That day she noticed a bluish fin sticking out from among the sharks and rays caught in his net. She later described her find as "the most beautiful fish I had ever seen, five feet long, and a pale mauve blue with iridescent silver markings."

  At that stage, she had no idea what the fish was. Her find, which was identified by a Mr Smith in 1939, a self-taught ichthyologist, was one of the most exciting zoological discoveries of the twentieth century.

- The ***male Sea Catfish*** goes one better than the Seahorse which carries the eggs in its belly sac. He incubates the female's eggs in his mouth and also allows his young to hide there until they are able to fend for themselves. All this time, he doesn't eat in case he swallows his own children!

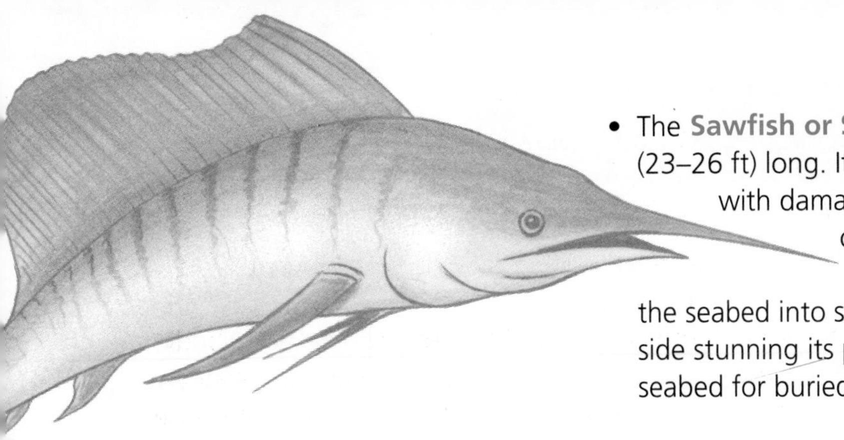

- The **Sawfish or Swordfish** is about 7–8 m long (23–26 ft) long. It has a saw-like blade instead of a snout, with damaging tooth-shaped scales on either side of it. The blade alone can be up to 2 m (6.8 ft) long. The Sawfish swims from the seabed into shoals of fish and lashes out from side to side stunning its prey. It also uses the blade to dig into the seabed for buried crustaceans.

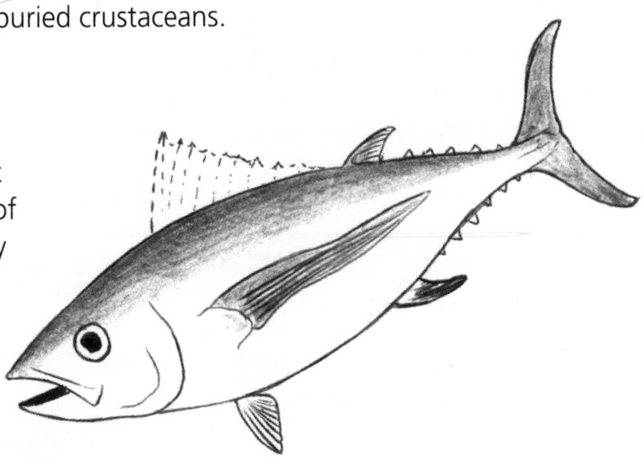

- The **Albacore** is a member of the tuna family. It is a mystery fish because no-one knows exactly where it breeds or where it raises its young. They are capable of covering huge distances in a single day. Suddenly they appear in great numbers off Japan's shores and are gone as quickly. To minimise resistance in water, the dorsal fin slips down into a slot in their back when swimming at speed.

- When sand or grit gets inside an ***oyster's shell,*** the oyster protects its soft body by coating the intruder with layers of mother-of-pearl. Gradually the layers build one on top of the other and a pearl forms. As far back as the 13th century the Chinese developed a method for producing artificial pearls.

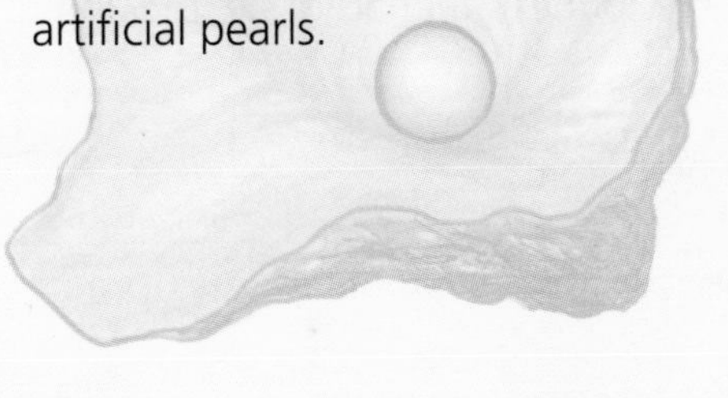

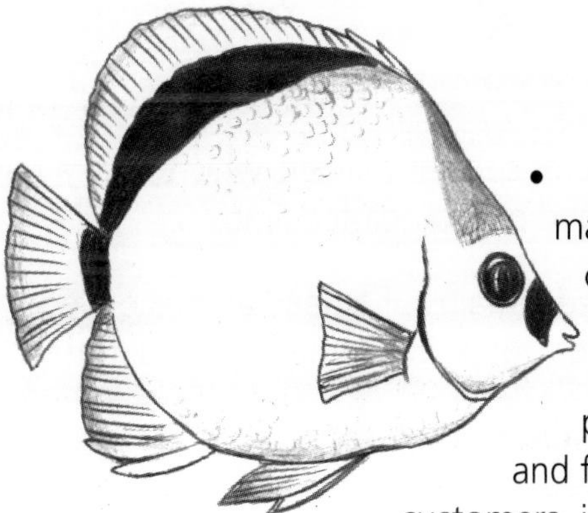

- **Barber Fish** make their living by cleaning other fish. They eat through the dead skin, parasites, bacteria and fungi of their customers, just as a barber cleans away facial hair on a person. They work in groups and advertise their services by staying close to brightly coloured sea anemones. Fish in need of a clean, wash and brush up, form queues. The Barber Fish are rewarded with a meal – one that has been delivered!

- ***Electric rays can give off electric shocks*** of several hundred volts. They have special organs on their undersides that build up electricity. They use the shock pulses to stun their prey. After spending most of their time hiding under sand or mud waiting, for victims to arrive.

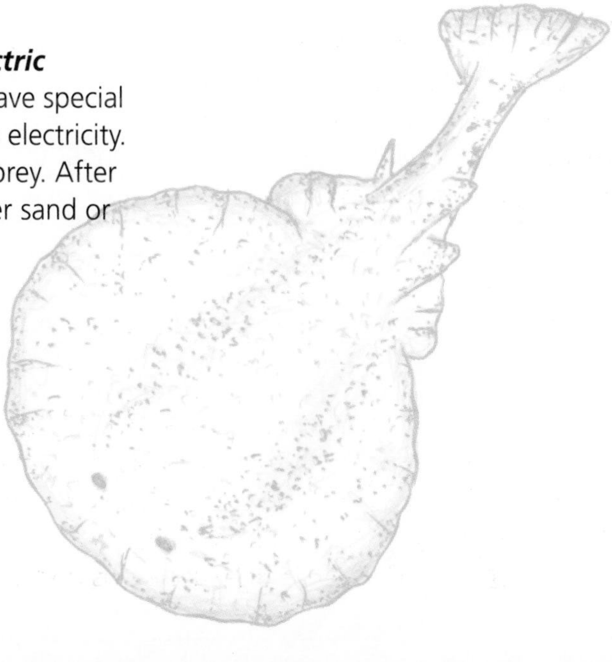

Glossary

You will need to remember these words whenever you discuss the world of sea creatures

Aerate To supply with air, or expose to the circulation of air.

Antennule A small antenna or similar organ. Especially small antennae on the head of a crustacean.

Baleen Made from keratin – the same type of compound in hair and fingernails. It is strong but flexible, and forms the filter plates in the mouth of the baleen whale.

Camouflage To disguise with blending colour that hides or deceives.

Carnivore A flesh-eating animal. Sperm whales are the largest carnivores on land or sea.

Cartilage A supple, gristly framework. Sharks have cartilage instead of bone, which makes them different from other fish. Our noses are built with cartilage.

Chromatophores Cells containing pigment granules enabling the organisms to change colour.

Crevice A narrow crack or opening.

Crustacean Creatures such as lobsters which have a hard outer skeleton (exoskeleton), a segmented body and paired, jointed limbs.

Debris The scattered remains of a dead animal or vegetable matter. Sea anemones eat debris from the sea floor. A word with a similar meaning is detritus.

Decibel A measure of the loudness of a sound. Blue Whales are the loudest animals of all, making a noise of 188 decibels.

Ferocious Extremely savage or fierce. A barracuda is a ferocious predator.

Flexibility Being able to bend easily. Belugas are the only members of the whale family with flexible necks.

Food chain A series of animals and plants, each depending on the next for food. The plants are usually at the bottom, with the fiercest predators at the top of the food chain.

Gelatinous A thick liquid.

Ichthyologist A zoologist who studies fishes.

Immobilise To prevent movement. In the world under the sea it's usually by stunning, or poisoning.

Incubate Sitting on eggs (or some other means) to provide heat. The male seahorse incubates its young in a belly sac-like pouch. Eggs will not hatch without proper incubation.

Invertebrate Lacking a backbone or spinal column. Jellyfish and squids are invertebrates.

Krill Very small marine crustaceans which are the main food for baleen and other whales.

Luminescent Something that is a source of light.

Manoeuvre To steer or travel in different directions as required.

Microscopic Can only be seen by using a microscope.

Mimicry Copying the behaviour, sound or appearance of another organism (animal or plant).

Mollusc Creature lacking a backbone (invertebrate) with a soft, non-segmented body usually encased in a shell. Certain octopuses and squid are molluscs but have no shells.

Nematocyst A stinging cell in the tentacles of many jellyfish.

Niche The particular spot where an animal or plant lives.

Palaeontologist A scientist who studies prehistoric life.

Parasite An organism that grows with, feeds and takes shelter from another organism. Examples of parasites are ticks, fleas and sea lice on fish.

Photophore A light-emitting organ showing as luminous parts of marine creatures.

Pinniped Walruses are pinnipeds – pinna meaning fin or wing, and pedis meaning foot.

Plankton The tiny organisms that float or drift in great numbers in the sea – food for many sea creatures, including baleen whales.

Predator An animal which kills and usually eats other animals.

Propulsion The action of driving or moving forward.

Regenerate To renew or produce again. A starfish can regenerate a ray (arm) when lost.

Secrete To manufacture and release. An octopus secretes black ink when alarmed.

Serrated Notched on the edge like a saw.

Shoal A school or group of fish of the same kind.

Symbiotic relationship An example is the relationship between the Clownfish that lives in the fronds of the tropical sea anemones. The fish protects the anemone from anemone-eating fish, and in turn the stings of the anemone protect the Clownfish from its predators. The two live in harmony.

Tentacle A long flexible "limb" such as that of the squid, used for feeling, grasping or moving.

Venom, venomous A poisonous secretion of an animal usually delivered by a bite or a sting.

Vulnerable One who has a weak defence against injury or harm.

Wean To train the young mammal to eat other food instead of suckling from its mother.

Abbreviations used

Metric	Imperial
Linear	
Centimetre = cm	Inch = in
Metre = m	Feet = ft
Kilometre = km	Mile = mi
Mass	
Gram = gm	Ounce = oz
Kilogram = kg	Pound = lb
Volume	
Litre = L	Pint = pt
Area	
Square metre = sq m	Square yard = sq yd
Square kilometre = sq km	Square mile = sq mi
Speed	
Kilometres per hour = kph	Miles per hour = mph

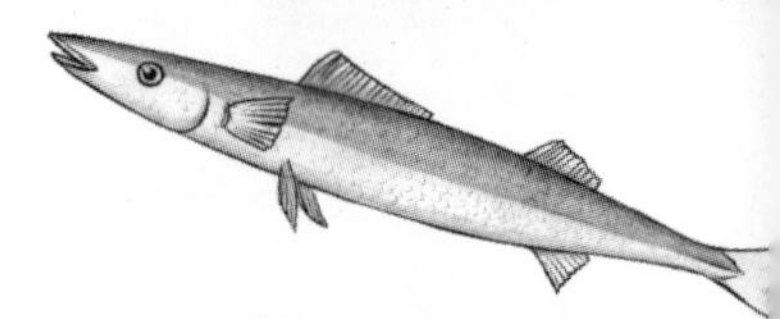

Wordsearch

T L M W S C L R N A S U S K L R R L
E O O A A L R A V U G E T R L O T P
N B L T R S E U O E T I T I C U L G
T S L E V C W M S A N A L L R E T R
A T U R O E O A R T S K E L E T O N
C E S O T N B E L C A A W A S L A E
L R C A E S N R T L N C S H A R K T
E S S V D E A N E M O N E F R E E F
S M O H G I K S T A T W L A I T S S
I N V E R T E B R A T E S C N S E A
A T R P M L E K T H A H A E F S H A
L T E A A C H I C N S L E P H W I L
G P W H E R B R O P A U S H L T L U
D S W B O A A T T A O E N A S G L P
A S E W H R K L E M X S T L P A O T
C O R A L N S E Y N Y S R O I L L L
P R E D A T O R S S G H H P S G Y T
K V T L S R L S C L E T O O E E H T
A N P S U C T I O N N H O D R Y R T
S L W S L T I F R L E T L P A G S S

ocean
lobsters
predators
habitat
venomous
sunlight
coral
fish

whales
shark
tentacles
swallow
anemone
plankton
krill
crustaceans

molluscs
skeleton
cephalopod
suction
reef
regenerate
paralyse

prey
salt
water
gills
breathe
oxygen
invertebrates

Look for words in all directions!